AF693930

HOW-TO SERIES

for the HR Professional

Linking Pay to Performance

Steven P. Seltz, CCP
Robert L. Heneman, Ph.D.

WorldatWork

About WorldatWork®

WorldatWork is the world's leading not-for-profit professional association dedicated to knowledge leadership in compensation, benefits and total rewards. Founded in 1955, WorldatWork focuses on human resources disciplines associated with attracting, retaining and motivating employees. Besides serving as the membership association of the professions, the WorldatWork family of organizations provides education, certification (Certified Compensation Professional — CCP®, Certified Benefits Professional — CBP™ and Global Remuneration Professional — GRP®), publications, knowledge resources, surveys, conferences, research and networking. WorldatWork Society of Certified Professionals and Alliance for Work-Life Progress (AWLP) are part of the WorldatWork family.

Any laws, regulations or other legal requirements noted in this publication are, to the best of the publisher's knowledge, accurate and current as of this book's publishing date. WorldatWork is providing this information with the understanding that WorldatWork is not engaged, directly or by implication, in rendering legal, accounting or other related professional services. You are urged to consult with an attorney, accountant or other qualified professional concerning your own specific situation and any questions that you may have related to that.

This book is published by WorldatWork. The interpretations, conclusions and recommendations in this book are those of the author and do not necessarily represent those of WorldatWork.

© 2004, 1993 WorldatWork
ISBN 1-57963-0154

No portion of this publication may be reproduced in any form without express written permission from WorldatWork.

WorldatWork.

The Professional Association for Compensation, Benefits and Total Rewards

WorldatWork
14040 N. Northsight Blvd., Scottsdale, AZ 85260
480/951-9191 Fax 480/483-8352
www.worldatwork.org

Publishing Manager: Dan Cafaro
Graphic Design: Mark Anthony Muñoz
Production Manager: Rebecca Williams Ficker
Staff Contributor: Bonnie Serino

Table of Contents

Overview

Merit pay, commonly called "pay for performance," is perhaps the most widely used means by which U.S. organizations determine employee pay increases. The purpose of merit pay is to reward employees for individual contributions and to encourage the best performance possible. In theory, if all employees operate at peak efficiency relative to their capabilities, the organization will thrive.

The logic behind merit pay is straightforward: If pay is made contingent upon performance, then employee motivation to achieve high performance is increased. (See Figure 1.) The key to merit pay is founded in three motivational theories:

- Reinforcement theory states that merit pay should motivate improved performance because the monetary consequences of good performance are made known – the better one's performance, the greater the pay increase will be.
- Expectancy theory states that merit pay should motivate improved performance because performance is instrumental to the attainment of a pay increase – improved effort to perform leads to increased pay.
- Equity theory states that merit pay should lead to improved performance because a pay raise is seen as a fair outcome for one's performance input – the more one contributes to the organization, the greater the pay increase.

A successful merit pay program will

- reward employees for achieving performance results and exhibiting behaviors aligned with the objectives of the organization, which often are linked directly to the strategic business plan and mission of the organization
- provide rewards commensurate with contributions (i.e., bigger pay increases for stronger performers)

FIGURE 1: LINKING PAY TO PERFORMANCE

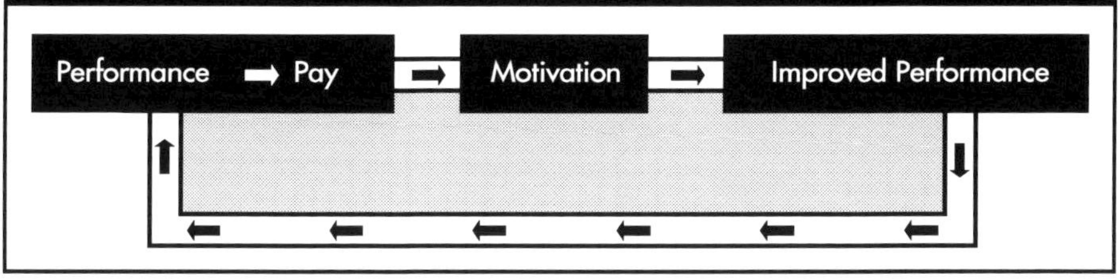

- be communicated easily to employees
- be understood readily by employees
- recognize "bottom line" considerations and the organization's ability to deliver pay increases
- be rational, structured and administered in a logical manner
- conform to legal requirements
- use a well-founded, credible means of evaluating performance
- conform with and support management philosophy.

As with most business programs, a merit pay plan should be planned carefully to achieve these goals. If an organization takes the time to design its merit pay plan carefully, it will establish a strong link between pay and performance.

1

Determining What to Reward

Before a merit pay plan can be designed, the first step is to determine

- what the organization values
- which types of individual employee contributions should be rewarded
- the organization's ability to pay
- the organization's ability and willingness to communicate the plan
- the organization's ability to administer the plan.

Some organizations make these determinations through the planning efforts of senior management, who refer to the overall business strategy and mission. Other organizations use a structured human resources planning effort, which relies on formal performance-planning and goal-setting activities. Other organizations make informal determinations. Some additional steps to getting started on a merit pay plan are listed in Figure 2.

Without a clear understanding of the organization's values and expectations, it is possible that employee contributions that are contrary to the organizational objectives will be rewarded. A successful plan requires that individual goals be aligned with the organization's:

- identity, which relates to whom the organization serves and what products and services are provided
- strategic plan, which relates to how the mission of the organization is accomplished
- objectives, which relate to what corporate goals have been established.

Once the link between individual and organizational objectives has been defined, merit pay can be used to align individual goals with those of the organization. When used properly, merit pay will reinforce the accomplishment of individual contributions that are in line with the identity, strategic plan and objectives of the organization.

Figure 2: GETTING STARTED ON A MERIT PLAN

Verify that key prerequisites are in place:

- top-management support
- an established performance-management system that is reliable, valid, fair,
- flexible and credible

Conduct research to verify that merit pay is appropriate and workable for the organization:

- review of prior merit pay theory and research
- collection of information on other employers' experiences with merit pay, focusing
- on those that are regarded as highly successful and highly unsuccessful as well as
- those that are similar in management style and organization
- evaluation of the effectiveness of the merit pay program by establishing a baseline of employee attitudes and perceptions about pay

Form an employee task force to oversee the development of the plan and ensure work-force "buy-in," with the following functions represented:

- line management
- HR professionals
- employees representing different "levels" in the organization
- nonexempt employees, if appropriate for the organization's culture

Merit pay also must be consistent with regard to the business environment of the organization. Business environment characteristics that support or detract from merit pay are shown in Figure 3.

Figure 3: HOW BUSINESS ENVIRONMENT CHARACTERISTICS RELATE TO MERIT PAY

MERIT PAY

SUPPORT →	← DETRACT
Little previous history of seniority-based raises	New or declining business
Emphasis on achievement	Unionized work force
Well-defined individual work outcomes	Highly task-interdependent employees
Top-management support	Egalitarian climate
Credible performance-management system	Inflationary environment
Cultural support for giving significant rewards to some and small or no rewards to others	Performance difficult to measure
Cultural support for communication of program tenets and characteristics	Low trust in management and human resources
High trust in management and human resources	

2

Documenting Performance Standards

The second step in developing a merit pay plan is to devise a system that establishes and evaluates performance against individual objectives. Performance standards, also known as performance goals or objectives, are written statements that help determine the extent to which employees have contributed to the mission of the organization. These standards establish the basis on which employee contributions are evaluated, and they define the expected level of performance. Various ratings systems can be used to describe how successful an employee has been in attaining objectives. Some examples of common performance standards are shown in Figure 4.

While establishing performance standards, determining which standards best meet an organization's needs is critical. Objective standards – such as

FIGURE 4: EXAMPLES OF COMMON PERFORMANCE STANDARDS

Quality

Demonstrates work quality by producing goods or services that meet or exceed preset, measurable standards (e.g., less than one defect per thousand items).

Quantity

Meets or exceeds specific production quotas within a given period of time.

Communications

Effectively expresses thoughts verbally, in writing and nonverbally. Listens attentively and makes productive use of acquired information. Gains agreement and acceptance of plans, ideas or activities being discussed while incorporating others' good suggestions.

Creativity and Innovations

Conceives, encourages, develops and applies imaginative concepts that improve operating procedures and efficiencies, or that make better use of company assets.

Interpersonal Teamwork

Works well with others toward the accomplishment of goals. Earns respect and trust. Makes a contribution to the team's achievements. Shows consideration for the feelings and needs of others.

Planning and Organizing

Defines and prioritizes objectives. Installs a thorough, appropriate plan of action. Establishes procedures to monitor progress toward task completion. Can manage multiple projects, priorities or deadlines to accomplish long- and short-term goals.

Problem Analysis

Identifies problems, secures relevant information and relates data from different sources to determine possible causes quality and quantity of work performed – should be assessed as well as more intangible, subjective aspects of the job such as teamwork, cooperation and customer service.

Documentation of work standards is an essential part of the performance-evaluation process. Typically, this is an annual event in which supervisors and subordinates discuss goals and objectives for the coming year while evaluating the prior year's performance. In some organizations, determining and documenting work standards is a cooperative effort between managers and employees. Other options include

- having managers determine objectives and then communicate them to the employee
- having employees present goals to their managers for discussion.

However work standards are established in an organization, obtaining employee "buy-in" is essential. If employees cannot comprehend the standards or accept their reasonableness, they are unlikely to perform in a manner that is consistent with the mission of the organization. To help ensure that employees accept and act upon performance standards, three actions should be taken:

- Emphasize results and behaviors rather than traits. Performance standards should reflect what the person produces (results) or what the person does (behaviors) rather than personality characteristics (traits). For example, it is better to measure the quality of performance by using a result such as "number of customer complaints" or a behavior such as "is always courteous to customers" than it is to use a trait such as "is nice to people."
- Employees should participate in setting standards. For employees to act upon performance standards, they must be committed to them, which means they need to have a sense of ownership in the process. When employees are given an opportunity to help establish performance goals and objectives, they are more likely to feel as if they "own" the process and to protect their ownership interests by meeting the standards.
- The standards should be flexible. It is the nature of work and organizations to be in a constant state of flux. Consequently, performance objectives and standards that are viable now may – due to influences outside the control of the employee – become obsolete. An organization should be willing to modify standards as shifting demands dictate.

3

Establishing a Merit Budget

A fundamental feature of any merit pay plan is an established budget that has been endorsed by management. Every year that a merit pay plan is in effect, the budget process should consist of two key activities:

- determining the size of the budget
- allocating funds to business units within the organization.

Determining Budget Size

Typically, salary-increase budgets will be established each year based upon many factors, including

- actual or anticipated organization financial results
- cost-of-living and/or inflation
- industry trends
- competitive factors such as retention rates and recruiting successes
- cost of labor and the competitive position of the organization's pay in the marketplace
- group (e.g., division or department) performance and needs.

In most organizations, it is common to obtain or develop salary budget surveys each year that show expected increase rates for similar employers. The American Compensation Association and many of the major compensation consulting groups conduct annual salary budget surveys and publicize their findings widely.

On the basis of survey information, and after taking into account the organization's financial situation, senior management ordinarily will approve a not-to-be-exceeded "bottom line" increase budget computed as a percentage of current payroll. During the past 10 years, merit-increase budgets have averaged approximately 5 percent annually.

Determining Budget Allocation

The next step in the budget process is to determine how funds are to be distributed to business units within the organization. A simple and commonly used method of allocating merit pay dollars is to use a uniform budget. Under this procedure, merit pay budgets are distributed to divisions or departments as a percentage of "eligible payroll," which is defined as the aggregate base salaries of all employees who are eligible to participate in the merit pay plan.

Eligibility may be driven by a calendar date. Some organizations include only those employees who have exceeded a minimum service requirement such as six months at the time of their expected date of increase. Other organizations will include all employees on payroll, but will prorate increases for those employees with a partial year of service.

Using the uniform-budget approach, every business unit in the organization shares proportionally in the amount of money available for salary increases. Figure 5 is an example of a uniform-budget allocation.

Figure 5: EXAMPLE OF A UNIFORM-BUDGET ALLOCATION

Department	Total Payroll Dollars	Merit Budget Percentage	Merit Budget Dollars
Finance	$2,450,500	4.0%	$98,020
Human Resources	$1,750,900	4.0%	$70,036
Marketing	$4,375,055	4.9%	$175,002
Production	$7,980,250	4.0%	$319,210
Totals	**$16,556,705**	**4.0%**	**$662,268**

Use of uniform budgets fails to take into account that some business groups are more or less successful than are others. Furthermore, in organizations with geographically dispersed business activities – some of which may be located in areas with different costs-of-living, inflationary or competitive pressures with respect to the work force – a uniform budget may be inappropriate.

To respond to differing achievement levels of the various business units or the need to pay different wages in certain locations, some organizations use a flexible-budget approach. The flexible-budget method introduces a level of complexity into the budget process that is not a factor in uniform merit budgets. Unlike uniform budgets, flexible budgets require sound measures of business-unit performance and geographic pay differences to distribute budget dollars. Many organizations are ill-prepared to track or calculate these differences accurately. Figure 6 is an example of a flexible-budget allocation. Note that budget percentages have been rounded.

Figure 6: EXAMPLE OF A FLEXIBLE-BUDGET ALLOCATION

Department	Total Payroll Dollars	Merit Budget Percentage	Merit Budget Dollars
Finance	$2,450,500	3.1%	$75,966
Human Resources	$1,750,900	3.6%	$63,032
Marketing	$4,375,055	4.3%	$188,099
Production	$7,980,250	4.2%	$335,171
Totals	**$16,556,705**	**4.0%**	**$662,268**

4

Setting Merit Pay Policy

The essential goal of a merit pay plan is to link pay to performance that is consistent with the mission of the organization. To cement this link, pay increases must vary according to the level of an employee's contributions and efforts. There are two required conditions:

- Variations in employee performance must be measurable and measured.
- Managers must be provided with the necessary "tools" to determine the appropriate rewards.

These tools are to be found in the established guidelines or policies that govern pay increases as well as in the process for implementing these guidelines.

Policy Decisions

Key factors in creating a merit pay policy are the size, timing and delivery of merit increases.

If differences among pay increases are deemed by recipients to be trivial, the merit pay program will be undermined because employees will not be motivated to improve their performance.

Size: Absolute vs. Relative

The size of pay increases is a critical component in merit pay programs. Two conditions are necessary to motivate employees most effectively to meet and exceed performance standards for their positions:

- The absolute size of the merit increase must be significant enough to make a noticeable difference to employees (e.g., the increase must not be so trivial as to be deemed inconsequential).

- The relative size of the increase must be significant enough that real differences in performance are recognized by meaningful differences in rewards.

A successful merit pay program will ensure that increases awarded to the "best" contributors will be substantially greater than increases awarded to average or less-than-average performers. If differences among pay increases are deemed by recipients to be trivial, the merit pay program will be undermined because employees will not be motivated to improve their performance. For example, a merit pay program that provides additional 2-percent increases for "exceptional" performers is likely to be perceived by employees as not providing significantly different rewards. However, a merit pay program that offers an additional 4- or 5-percent increase for exceptional performance is more likely to alter employee behavior and be motivational.

The anniversary-date approach focuses the performance evaluation and increase on an individual employee, ideally leading the employee to believe the process is focused specifically on him or her.

Timing: Anniversary vs. Common Review

Another issue that must be addressed is the date merit-increase decisions are made. Survey data suggest that common review dates are used by almost two-thirds of organizations while one-third stagger increases, mostly by providing them on anniversary dates.

Using an anniversary-date approach spreads the administrative burden (tasks such as completing performance reviews, making increase decisions and processing pay increases) throughout the year for managers and human resources staff. Payroll increases also are staggered, reducing the financial impact that accompanies a single jump in salaries. Also, the anniversary-date approach focuses the performance evaluation and increase on an individual employee, ideally leading the employee to believe the process is focused specifically on him or her.

A disadvantage of an anniversary-date approach is that relative performance (e.g., comparative appraisal ratings), which often is a basis for merit-increase decisions, may be hard to judge, particularly if performance is evaluated at different times for all employees. Another disadvantage becomes evident when conservative budget management accentuates the natural tendency of many managers to "save" money until year-end. When this occurs, employee increases at the beginning of the year may be smaller than increases at the end of the year, and the result may be to penalize some employees unfairly.

A common (i.e., annual) review date consolidates the administrative burden for management and human resources, and increases can become part of the yearly budgeting process. Furthermore, because increases for all employees are determined at the same time, appraisal ratings for all employees can be collected and relative performance can be factored into the decision more easily. If the merit budget is based on business-unit performance, the linkage among business-unit performance, individual performance and merit increases can be clearer with a common date.

Disadvantages of common review dates are that the workload may be onerous if the timing of the salary-increase program coincides with other major efforts (such as year-end financial closings, open enrollment for benefits, and departmental budgeting), and cash-flow implications for the organization may be extreme when all increases occur simultaneously.

Neither an anniversary-date nor a common-review-date approach is the "right" way to administer merit increases. The approach should be determined by several factors, including availability of performance data for employees and for organizational units, and the availability of management and human resources.

In organizations where budgets for salary increases are allocated based upon organizational performance during a fixed period of time, employee and departmental performance also may be evaluated during the same time period. In such cases, a common review date may make sense. Anniversary-date increases may be more appropriate in organizations that stagger appraisals of performance or that permit little or no increase-budget variability among departments, and in organizations that want to emphasize the individual's performance against absolute standards instead of emphasizing relative performance.

Another issue is whether to permit variability in time between increases in the pay program. In some organizations, the time between increases is not uniform for all employees; rather, performance differences are reflected not only in the size of increase but also in frequency. Excellent performance may be rewarded with larger and more frequent rewards. For example, top contributors might receive relatively large pay raises every six to nine months, while average performers might wait 12 to 15 months for a lesser increase.

In organizations where employees are at or over the maximum of their grade, but are not permitted to receive increases, lump sums provide a mechanism to continue to reward and motivate strong but highly paid contributors.

Delivery: Base vs. Lump Sum

Under traditional merit pay plans, merit increases are built into employees' salaries for as long as they remain with the organization. Hence, the increases are permanent, and their values are compounded over time as additional increases are granted.

One alternative to base-salary increases that recently has become more popular is the use of lump-sum increases. Lump-sum increases are one-time payments made in lieu of a traditional base-pay increase, and they typically are delivered annually via the merit pay program. Similar to a "bonus" payment, a lump sum must be re-earned each year based on performance – it is not built into base salary. Often, lump-sum payments are provided to employees who are near, at or over the maximum of their salary range (often called "red circle" employees).

The advantages of lump-sum increases for the organization are clear:

While retaining a pay-for-performance relationship, payroll costs over time are lessened because of the lack of a compounding effect. Also, the "sanctity" of pay ranges is protected because the number of red-circle employees can be controlled. In organizations where employees are at or over the maximum of their grade, but are not permitted to receive increases, lump sums provide a mechanism to continue to reward and motivate strong but highly paid contributors.

There are fewer advantages of lump sums for the employee, though receiving the annual increase at once rather than having it paid out over 12 months – as is the case with a base-pay increase – may appeal to some. For long-term and highly paid employees, who may be near the top of their salary range with no room to grow, lump sums provide a means to continue to receive rewards.

Frequently, lump sums issued in lieu of merit increases concern many employees because their gross (base) pay will be less over time. Longer-term employees approaching retirement typically exhibit the most concern. Many employers allay this concern by counting lump-sum awards toward final average-earnings pension calculations. Similarly, such payments often are tied to benefits. For example, benefits such as life insurance, which are linked to salary, will reflect lump-sum payments in addition to base salary. This solution addresses a number of issues:

- The motivational link between performance and reward can be maintained.
- The organization reaps the benefit that lump-sum payments provide in controlling total compensation costs.
- Employee benefits entitlements are not seriously reduced.

However, caution should be used with this approach because employees may react negatively when their base salaries do not change or grow relatively slowly over time.

Policy Implementation

A merit pay policy answers the following questions about salary increases: How much? When? How? How frequently? These decisions can be summarized in a simple compensation tool called a merit pay matrix.

A merit pay matrix details the amount and timing of increases for various levels of performance at various locations in the pay grade. A merit pay matrix may be interpreted as an operational statement of an organization's pay-for-performance theory or policy. It spells out the contingency between pay and performance in specific terms.

Merit pay matrices range from the simple to the complex, depending on the number of variables upon which pay is made contingent. Generally, there are three alternatives for issuing merit increases:

- based only on performance
- based on performance and position in range
- based on performance and position in range using variable timing.

Performance

This method, which uses the simplest form of merit matrix (Figure 7), is most common in organizations without well-defined salary grade structures. Pay increases are granted based solely on performance, resulting in top performers receiving bigger increases than lower performers. Typically, salary increases are calculated as a percentage increase in base pay.

Basing merit increases on performance alone ignores internal pay comparisons. Within a performance class, higher-paid employees receive greater absolute increases, even though the merit percentage reward is the same. This has the effect of perpetuating pay inequities that might exist, and it may reward long-tenured and/or highly paid employees disproportionately.

An alternative is to calculate merit increases as a percentage of the employee's salary-grade midpoint rather than of base pay. This approach provides larger relative dollar increases to employees within a performance class who are paid lower

FIGURE 7: LINKING MERIT INCREASES TO BASE PAY

Performance Rating	Fixed Increase Amount	Discretionary Increase Amount
Outstanding	8%	6-10%
Consistently Exceeds Standards	5%	4-6%
Meets Standards	3%	2-4%
Does Not Fully Meet Standards	0%	0-2%

Note: The matrices displayed in Figures 7, 9 and 10 use ranges of increases rather than single percentages. This provides for more managerial discretion in awarding increases, and it more closely links pay and performance. In some companies, however, each cell of the matrix is occupied by only a single number.

in their salary range than it does for employees who are high in their range (See Figure 8 on page 16). Over time, inequities in salaries of employees in the same salary grade will be reduced as lower-paid employees are accelerated toward midpoint and higher-paid employees are "slowed down." This method reduces some of the bias toward long-term/highly paid employees that may be inherent in a performance-only merit matrix.

The advantage of either approach to calculating merit increases based only on performance is that the method is

- simple to budget
- easy to administer
- straightforward to communicate.

FIGURE 8: LINKING MERIT INCREASES TO SALARY-GRADE MIDPOINTS

Increase as a Percentage of Base Pay

Employee	Current Pay Rate	Increase Percentage	Increase Dollars
A	$25,000	4.0%	$1,000
B	$35,000	4.0%	$1,400
C	$45,000	4.0%	$1,800

Increase as a Percentage of $35,000 Midpoint

Employee	Current Pay Rate	Increase Percentage of Midpoint	Increase Dollars	Effective Increase Percentage
A	$25,000	4.0%	$1,400	5.6%
B	$35,000	4.0%	$1,400	4.0%
C	$45,000	4.0%	$1,400	3.1%

Performance and Position in Range

Organizations with more traditional and/or sophisticated grading structures commonly introduce the practice of basing increases on both performance and position in range, which is commonly defined by quartiles, or, if greater precision is required, by compa-ratios. This practice is based on the concept that the midpoint represents a "competitive" or "fair" wage for a particular set of skills in the marketplace, and that, over time, employees with a similar level of sustained performance should be paid an equivalent amount. Thus, a merit-increase guide chart similar to Figure 9 will cause employees with the same performance to converge, over time, on a target point (typically the midpoint) by awarding bigger increases to employees lower in their range and smaller increases to employees higher in the range.

This merit-matrix approach has several advantages:

- The tendency is reduced to perpetuate tenure-based pay inequities and to continue to "overpay" (relative to market) highly paid employees.
- The approach is more likely to be deemed "fair" by the work force because, over time, employees with similar performance in the same salary grade will tend to be paid comparably.

Basing merit increases on both performance and position in range introduces a level of complexity into the process not found in the simpler performance-only model. It is more difficult to administer and communicate.

FIGURE 9: LINKING MERIT INCREASES TO PERFORMANCE AND POSITION IN RANGE

	Position in Range Before Increase			
Performance Rating	**1st Quartile or Below**	**2nd Quartile**	**3rd Quartile**	**4th Quartile**
Outstanding	8-9%	6-7%	4-5%	3-4%
Consistently Exceeds Standards	6-7%	4-5%	3-4%	2-3%
Meets Standards	4-5%	3-4%	2-3%	X
Does Not Fully Meet Standards	0-2%	X	X	X

Performance and Position in Range Using Variable Timing

A more complex model for administering merit increases involves the concept of variable timing. The guide chart shown in Figure 10 demonstrates how the size and frequency of increase can be varied based upon performance and position in range. In this model, top performers receive bigger and more frequent increases, while average and below-average contributors wait longer for smaller increases.

There are several advantages to this approach:

- Top performers will receive bigger rewards with greater frequency, yielding significant increases due to the compounding effect.
- During times of tight budgets, rather than issuing "below market"

increases at regular intervals, "normal" increases can be granted at moderately delayed intervals. For example, rather than granting a 3.5-percent increase at 12 months, an organization may prefer to grant a 4.7-percent increase at 16 months.

The disadvantages of variable timing are

- It is much more complicated to administer.
- It is difficult to track and maintain budgets.
- It is difficult to monitor the consistency of application throughout the year.
- It is more complex to communicate.

Figure 10: LINKING MERIT INCREASES TO POSITION IN RANGE USING VARIABLE TIMING

	Position in Range Before Increase			
Performance Rating	**1st Quartile or Below**	**2nd Quartile**	**3rd Quartile**	**4th Quartile**
Outstanding	8-9% 6-9 Months	6-7% 9-12 Months	4-5% 10-12 Months	3-4% 12-15 Months
Consistently Exceeds Standards	6-7% 8-10 Months	4-5% 10-12 Months	3-4% 12-15 Months	2-3% 15-18 Months
Meets Standards	4-5% 9-12 Months	3-4% 12-15 Months	2-3% 15-18 Months	X
Does Not Fully Meet Standards	0-2% 12-15 Months	X	X	X

A successful merit pay plan requires more than well-developed policy statements and a conceptually sound design. It also requires administrative processes and procedures that are logical and easily understood. Some of the administrative issues that should be given consideration to ensure that a policy is implemented as intended are communication, training and perceived fairness.

5

Managing a Merit Pay Plan

Honest, open communication between management, human resources and employees serves as the means by which the messages of merit pay can be conveyed and reinforced.

Communication

The merit pay "equation" is simple: Significant performance efforts yield significant rewards, which in turn motivate significant performance efforts. However, this equation relies on trust to enforce the contract between employees and the organization. Employees must trust the organization to fulfill its commitment that today's efforts will be compensated fairly tomorrow, and the organization must trust that employees will be motivated by performance-based rewards.

As in any relationship, trust can be promoted through openness and candor or thwarted through secrecy and obfuscation. Honest, open communication between management, human resources and employees serves as the means by which the messages of merit pay can be conveyed and reinforced.

Traditionally, many organizations have been unwilling to share much of their compensation-related data. Usually, these organizations have the mistaken belief that employees neither want nor need to know about such matters, and that providing "too much" information to employees somehow reduces management's ability to exercise flexibility and discretion.

In recent years, many organizations have come to believe that no matter how carefully designed or well-founded a compensation program might be in theory, success requires adequate communication. Thorough communication permits employees to test the validity of the organization's promises while conveying to them that the organization has nothing to hide. It also establishes opportunities for dialogue on issues of critical importance, enhances credibility by obtaining employee buy-in and promotes overall trust.

A successful communication program requires a careful balance between an insufficient amount of information and too much information. Management should release enough information about the plan to demonstrate its faith in the process, but not so much information that its ability to exercise managerial discretion is impeded. Employees should be provided enough information about the merit pay plan to serve as a performance motivator without breaching their right to privacy.

How much to communicate to employees will be influenced by many factors, including the organization's culture, management's willingness to share information that traditionally may have been confidential, and the readiness and ability of human resources to support the communications effort. Some of the key elements often introduced in a comprehensive communications program are

- general information about the performance-appraisal program and process
- general information about the organization's compensation program (e.g., how pay is determined, how jobs are evaluated and what the salary ranges are)
- more specific information about the merit pay program (e.g., salary-increase budgets, performance-rating distributions and merit matrices)
- size of an individual's increase, minimum and maximum raises granted, and average size of merit increases.

Training

Implementation of a successful merit pay program requires managers to make two key sets of decisions:

- evaluation of performance
- allocation of increase awards.

An accurate, reliable and credible performance-appraisal program is the foundation of a successful merit pay program, and it is imperative that managers and supervisors be capable of evaluating employee behaviors and results objectively and critically.

The skills required to appraise performance, assess employee contributions and assign rewards are not intuitive. To ensure adequate interpretation and understanding of program requirements and consistent application of program tenets, training should be provided for all managers who are given the task of implementing the merit pay program. Training should include the following components:

- how to plan performance that links individual efforts and accomplishments to business plans and strategies
- how to measure and evaluate performance fairly and consistently
- how to provide feedback through intrinsic (e.g., coaching and praise) and extrinsic (e.g., pay increases and incentive payments) rewards
- how to use the merit matrix to allocate rewards
- how to communicate the assessment of performance and the allocation of rewards to employees.

Perception of Fairness

Program credibility is key to gaining a favorable response among employees to merit pay. Employees need to feel that increases and the process used to derive the increases are accurate and fair. To help ensure the perception of fairness, the merit pay program should incorporate the following tenets:

How Computer Technology Can Assist Administration

The most conceptually and theoretically sound merit pay program is burdensome and inefficient to administer. Consequently, anything that contributes to simplify planning and administration will help ensure the program's success.

Computer technology can assist in the management of the merit pay plan in a number of ways:

- Budget planning can be facilitated by generating different increase-matrix models, testing various options and deriving forecasts of the economic impact of alternatives.
- Data can be probed to evaluate the effectiveness, impact and equity of the merit pay plan. Increases, performance distribution and other factors can be analyzed by department, position, organization or individual.
- Employee records can be stored, monitored, and analyzed over time.
- Data can be managed to formulate cost projections based on salary-structure changes, the impact of inflation and other financial factors.
- Summary reporting can be streamlined for internal and external purposes, tedious administrative tasks and reporting efforts can be automated, and productivity can be improved by reducing the amount of time, labor and expense involved in managing the pay program.

- Relevant laws and regulations must be followed (e.g., Title VII, the Civil Rights Act, the Fair Labor Standards Act and various tax laws).
- Employees should participate in the setting of performance goals and standards, they should know what performance is expected of them, and they should be able to control the specific aspects of their performance on which their pay will be based.
- Employees should know and understand how the pay program works, and they should be encouraged to raise concerns, ask questions and seek clarification of their increases.
- An appeals process should be established to provide employees with an opportunity to discuss their performance evaluation and their increase with an authority other than their direct supervisor.

6

Evaluating a Merit Pay Plan

To ensure that a merit pay plan is operating as intended and is effective in meeting an organization's compensation needs, a systematic, post-implementation evaluation of the plan should be conducted regularly. This often-overlooked step is critical to the ultimate success and acceptance of the program. Many factors can be analyzed to assess plan effectiveness:

- employee satisfaction with the pay program
- employee job satisfaction
- employee perception that pay is based on performance
- employee acceptance of and trust in the performance-appraisal process
- employee trust in management
- employee and organizational performance (e.g., productivity improvements)
- employee commitment to the organization as demonstrated through reduced turnover and absenteeism rates
- correlation between actual performance ratings and actual merit increases.

Measurement of these success factors before and after implementation of a merit pay plan is likely to yield the most meaningful information, and it can be accomplished through various means: controlled empirical studies, employee-attitude surveys, focus-group discussions, and management and employee anecdotal feedback. Employee attitudes and perceptions ideally should be evaluated by collecting survey data from employees before the introduction of a merit pay plan and again after the program has been introduced and employees have received their first merit increases.

Some organizations attempt to gauge the success of newly introduced merit pay plans by measuring productivity and/or performance improvements over time and then correlating that information with appraisal ratings and salary increases. Also, turnover and absenteeism rates can be tracked and correlated with performance and salary increases. These data could be used to modify development of the plan, but it should be remembered that many other factors, including industry and economic trends, also may affect these factors. For example, high unemployment rates will tend to drive down turnover rates, regardless of employee satisfaction with corporate pay plans.

Because employee perception of fairness is so important in determining the success of the merit pay program, one analysis that should be performed is to test how accurately, fairly and consistently the program has been administered throughout the organization. Some common employee questions that need to be addressed to demonstrate the fairness of a merit pay plan are

- Does where or for whom you work mean more than how well you perform? Do some departments rate employee performance disproportionately high, and are some supervisors unfairly critical while others are unreasonably generous?
- Are all employees afforded a relatively equal opportunity for high performance ratings and commensurate increases? Is the plan free from racial, gender and age bias?

Merit Pay Advantages and Disadvantages

While merit pay remains a popular means of determining pay increases, the potential drawbacks of the approach should be clear before implementation. Once these drawbacks are recognized, an organization can appreciate the advantages of merit pay and how it will improve employee perceptions of work and rewards (Figure 11).

As is the case with any reward system, merit pay must be compatible with an organization's culture and philosophy if it is going to be effective. For example, merit pay will not work for an organization that values tenure over performance. Also, merit pay may be inappropriate for a growing number of organizations that are trying to emphasize group performance instead of individual performance. By rewarding individuals, merit pay can help undermine the cooperation and interdependency that are needed in a team environment. However, it may be possible to preserve the best elements of a team environment while rewarding the highest-performing individuals by integrating group-based incentives with some form of merit pay system.

Merit pay will not work unless an organization has a sound system of measuring individual employee performance that is accepted by the work

FIGURE 11: ADVANTAGES AND DISADVANTAGES OF MERIT PAY

MERIT PAY

ADVANTAGES →	← DISADVANTAGES
Helps improve employee satisfaction with work and pay as well as individual performance	Rewards individual performance, not group performance
Rewards performance rather than seniority or skills	Depends highly upon a sound performance-appraisal system
Clarifies performance expectations	Clashes with organizational emphasis on tenure
Attracts and retains highly motivated employees	De-emphasizes intrinsic work rewards while possibly discouraging "average" and "below average" performers

force. Even if a good performance-appraisal system exists, merit pay may be discouraging to "average" or "below average" employees, who typically will fail to qualify for high pay raises. By tightly linking pay and performance, merit pay also can de-emphasize the intrinsic rewards and satisfaction gained simply from doing a job.

By linking a merit pay program with a sound communications strategy, an organization can clarify its performance expectations and create an atmosphere of trust between employees and management. This atmosphere tends to increase overall employee satisfaction with work and pay, and it is likely to lead to improved individual performance.

The main reason an organization chooses a reward system is to enhance its competitiveness, productivity and bottom-line results. Positive financial results are more likely when an organization places emphasis on employee performance instead of tenure, and highly motivated employees are more likely to be attracted and retained when their efforts are rewarded regularly. A merit pay system can help ensure that an organization's rewards policy fits the performance-based philosophy it needs to survive and prosper.